Python:

A Beginners' Guide to Python Programming to automate the boring tasks and learn coding fast

Table of Contents

Disclaimer 4

About the author 6

Introduction to Python 8

    Why is Python So Popular? 10

    Outstanding Features of Python 12

Installing Python on Various Platforms 18

    Installing Python on Microsoft Windows 18

    Running Few Programs on Your New Install 23

    Installing Python on Linux 27

    Installing Python on Mac OS X 28

Your First Steps towards Learning Python 31

    Using the Interpreter Prompt 32

    Using an Editor in Python 34

    Using the Source File 43

Python Basics for Beginners 47

    Comments 48

    Literal Constants 49

    Quotes 50

    The Format Method 53

    Escape Sequences 56

    A Program to Practice 58

Various Operators & Expressions Used in Python 64
    Operators in Python.. 66
    Operator Precedence Table for Python 77
    Expressions... 80
Control Flow Statements in Python 85
    The If Statement 86
    The While Statement 94
    The Recursive Function in Python................................... 100
    Functions... 102
Files & Storing Persistently in Python 105
    Storing Persistently in Python.. 110
    cPickle 115
Sample Programs for Practice... 117
    Exercise for You.. 130
Conclusion... 133

Disclaimer

Copyright © 2016

All Rights Reserved

No part of this eBook can be transmitted or reproduced in any form including print, electronic, photocopying, scanning, mechanical or recording without prior written permission from the author.

While the author has taken the utmost effort to ensure the accuracy of the written content, all readers are advised to follow information mentioned herein at their own risk. The author cannot be held responsible for any personal or commercial damage caused by information. All readers are encouraged to seek professional advice when needed.

About the author

John Slavio is a programmer who is passionate about the reach of the internet and the interaction of the internet with daily devices. He has automated several home devices to make them 'smart' and connect them to high speed internet. His passions involve computer security, iOT, hardware programming and blogging. Below is a list of his books:

John Slavio Special

Introduction to Python

At some point in time or the other, a lot of people get the urge to start programming. However, most of them never make a move towards this field because computer programming sounds scary. Well, you should know that it's a lot easier than it sounds. All it takes is the correct selection of programming languages to begin your journey as a computer programmer. Programming languages are made to make human life easier. These languages make programs that increase the overall productivity, communication, and efficiency of work. Of the many programming languages, Python is an all-time favorite. This is because Python is one of those rare languages that is both simple, and powerful. Python has, indeed, everything that you require to make a new program. If you should see someone writing a program in Python, you would be surprised at how easy it is to find the solution for a problem in that language.

Python is defined as "a powerful, yet simple programming language. Python consists of high-level data structures. Besides, it also has an effective procedure for object-oriented programming.

Python's neat syntax, along with its interpreted nature and dynamic typing makes it the perfect programming language for quick application development."

In simple terms, python is an open source, high-level programming language. It was developed by Guido van Rossum in the 1980s and is presently administered by Python Software Foundation.

Why is Python So Popular?

As I indicated earlier, programming languages are created to make human life better. However, many languages become obsolete over time when they fail to keep pace with human needs and expectations as technology changes. Python has proven its worth over the years in both business and industrial use. Unlike other languages, it did not go obsolete, but in fact, has increased its use with time. Python is broadly used in making web applications, GUIs (Graphical user interface), games, etc., and writing and reading the codes in the program are as simple as reading regular English sentences. Programs written in Python are required to be processed before running, as they are not written in machine readable language. Once you learn the basics of Python (as you are a beginner), you will find it easy to move on to the advanced features that the program offers to programmers. Below are some features of Python that make it a favorite.

Outstanding Features of Python

Python is simple

Python is a simple and minimalistic programming language. It is problem-oriented rather than focused on machine language. That is, the coding language is written in simple English, giving the user enough time to focus on the problem and find a solution, rather than worry about the mistakes that he could make while writing the program.

Python is easy to learn

Python has the simplest syntax that you can find in any programming language; therefore, even a beginner can learn it easily in a short time.

Python is free and open sourced

Python is the best example of Free/Libre and Open Source Software (FLOSS). This is its strength as FLOSS gives the Python community the opportunity to make changes to the program, to contribute to its development, to distribute it, to make use of it in creating new software, and to read its source. The Python community works constantly without any worry about legal issues to ensure that it gets better over time.

Python is portable

Because it is an open source software, Python has been ported to many platforms, thus making it the most portable programming language available. All your Python programs can work on any of the following platforms without the need for changes to the Python file. You can use it on GNU/Linux, Windows,

Macintosh, FreeBSD, OS/2, AROS, Amiga, Solaris, AS/400, OS/390, BeOS, z/OS, QNX, Palm OS, VMS, Acorn RISC OS, Psion, VxWorks, Sharp Zaurus, PlayStation, PocketPC and Windows CE.

We will discuss how you can install Python on various platforms in the next chapter of this book.

Python is embeddable

The user can easily embed Python with C or C++ programs.

Python is both procedure oriented and object oriented

When using Python, a user can choose to use it as a procedure oriented language or object oriented language. In a procedure oriented language, the program is built around functions or procedures that are reusable pieces of software. While in object oriented language, the program is built around objects which combines the functionality and data. When compared to vast languages like Java or C++, Python is the simplest object oriented program available today.

Can you now understand why you should learn Python as your first programming language? If yes, then let's continue to the next chapter.

Installing Python on Various Platforms

As mentioned earlier, Python is available for various platforms because of it being an open source software. Let's discuss how you can install it on your system. Select your Operating System (OS) below and follow the steps to install it.

Installing Python on Microsoft Windows

The first step in installing Python on Microsoft Windows is to download the software. You can download it from Python's official website at the link below:

https://www.python.org/downloads/windows/

- Once you click on the version that you want to install, you will go to the next page where you can see all the recent changes and features that were added to it. Scroll down on that page to reach the Files section.
- From the Files section, you can download Python. Ensure that you select the correct Operating System for your download.

- Once the download is complete, open the downloaded file and follow the on-screen instructions.
- Once you are on the Customize screen, scroll down to find the "Add Python.exe to Path", there you will see a red 'x'. Click on this icon and select the "Will be installed on local hard drive" option, then click "Next."
- Now the command prompt will start which will download and install Python on your machine.
- After the installation is complete, you can exit the setup by clicking on the 'finish' button.

- The next step is adding Python to system path variables. If you downloaded version 3 for Python, this step is not required. However, if you installed Python version 2, you will have to complete this step. Doing this step ensures that Python can work perfectly and can run all its scripts on the Windows machine without creating any conflict with your operating system.
- Start by opening your Start menu on your machine and then typing the word 'Environment'. Then select

'Edit the system environment variables.'

- After that, system properties will open in front of you in a new window. At the bottom of this window, you will see a button named 'Environment Variables...' proceed by clicking on this button.
- Once the window with Environment Variables opens, you will see the bottom section named as 'System Variables'. Create the new variable by clicking on 'New'.
- Now, a new window will open with options to enter the name and

variable value. In the name box, enter the name of your script, and in the value slot, enter the code. The code is given below – 'C:\Python27\;C:\Python27\Scripts;'

- The above code may change depending on the Python version that you installed. Now save all the changes and you are done.

Running Few Programs on Your New Install

In the previous section, you learned how to install Python on your Microsoft Windows machine. Now, it is time to test your Python installation by running some basic programs in it for a start. Let's start with opening Python on your Windows machine.

- To open Python, click on Start and type "python". Two or three options will appear. Select "IDLE (PythonGUI)".
- Once Python starts, you can test its functioning by using a simple Print directive as explained in this step. This directive which print whatever you type on your screen as it is. Begin by typing your first directive similar to the one in the below

image, you can change the green line to whatever you want.

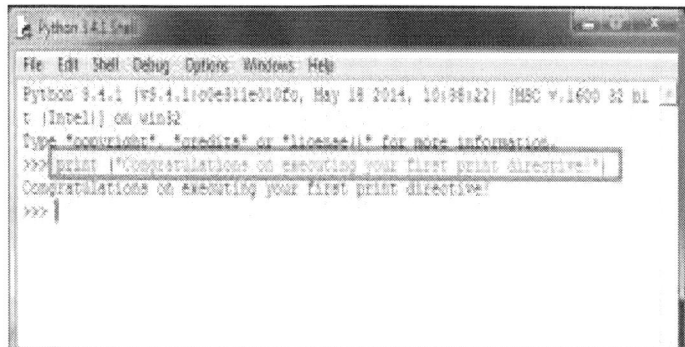

- If you installed Python 3 on your machine, then the symbol >>> must have caught your attention already. This symbol indicates that you can perform simple mathematics in Python without needing any sort of directive. Try doing some of the simple mathematics questions like the ones shown in this image -

- If you could perform these simple operations in Python, then it is time to congratulate yourself. Yes, you have finally installed your programming language on your Microsoft Windows machine.

Now, let's look at other operating systems available and learn how you can install Python on them.

Installing Python on Linux

You might not know this but for GNU or Linux users, Python comes already installed by default. To check if you have Python already installed, simply open the terminal application on your Linux machine, or press **Alt+F2** from your keyboard and then enter gnome-terminal. Now type the following command to continue checking for Python and which version is installed:

$ python –V

It may give you a result like this –

Python 2.7.6

Installing Python on Mac OS X

Just like GNU or Linux, Mac OS X also comes with Python already installed by default.

- To check this, simply press the **Command and Space** keys together from your keyboard to open the Spotlight search.
- Then type 'Terminal' and press the Enter key.
- Now, run Python and verify your install to ensure that there are no errors.

So, that's it for installing Python on your machine. From now on in this book, it will be taken for granted that you have Python already installed on your system. That being the case, why don't we get used to Python by learning few basics about it in the next chapter?

Your First Steps towards Learning Python

Earlier, we ran a simple print operation in Python in Microsoft Windows to check if it was properly installed. We will again run a simple, yet traditional print command in Python to understand how to run, save and write programs using Python.

When using Python for writing a program, you have two choices. One is to use the Interpreter Prompt to write your program and the other by using a Source file.

Using the Interpreter Prompt

Begin by opening the Python terminal as we did when we were installing it on your machine. Once you start the Python program, you will see the >>> symbol on your screen. Start typing in front of the symbol. This icon is called Interpreter Prompt. Now, let's learn how to use this interpreter prompt by writing another simple print program as we did earlier in this chapter.

Next to >>> symbol, type: *print('Hello World')* and press the enter key. You will now see Hello World printed on your screen.

You will notice in the above program that Python returns the result for your input as soon as you hit the enter key. What you just typed in Python is known as single Python statement. At the end, you will notice that the >>> symbol appears again, which means that you can now write another line in Python.

After writing a program, if you need to quit the interpreter prompt, you can do so by using **Ctrl+D** from your keyboard, or by entering exit() if you are using Linux or Mac machine. However, if you are on Microsoft Windows on your machine, use **Ctrl+Z** from your keyboard and then hit **Enter key** to exit the interpreter prompt.

Using an Editor in Python

The interpreter prompt is okay for writing the programs in Python. However, we cannot keep typing our program again and again whenever we want to execute it. To save time, we need to save our programs in files from where we can run them as many times as we want when they are needed. The files used to save the programs are known as source files. For creating the Python source file, an editor software is required for Python where we can write our program and save it.

Selecting the correct editor program is important as the better the program is, the easier it will be for you to write in it, therefore saving you a lot of time. One of the most basic requirements of a good editor is syntax highlighting, thus coloring all the different parts or lines of your program. Because of this, you can visualize the program running while seeing it on your screen.

As a beginner, I recommend that you go for the PyCharm editor software. This software is available for Microsoft Windows, GNU/Linux, and Mac OS X. The program is available for free on its official website. For the rest of the process of using an editor, I take it that you have downloaded and installed PyCharm on your machine. Now, once you open PyCharm, you will see a screen like this –

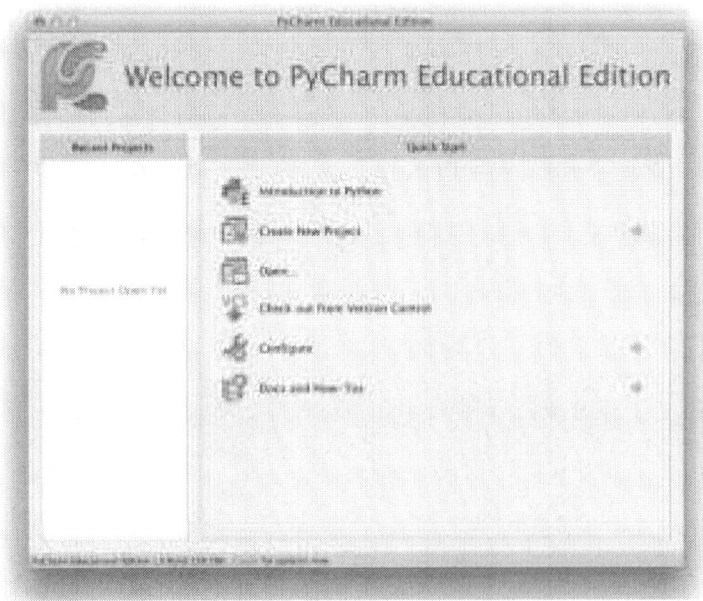

Click on 'Create New Project' and another screen will open up with 2 options as shown below.

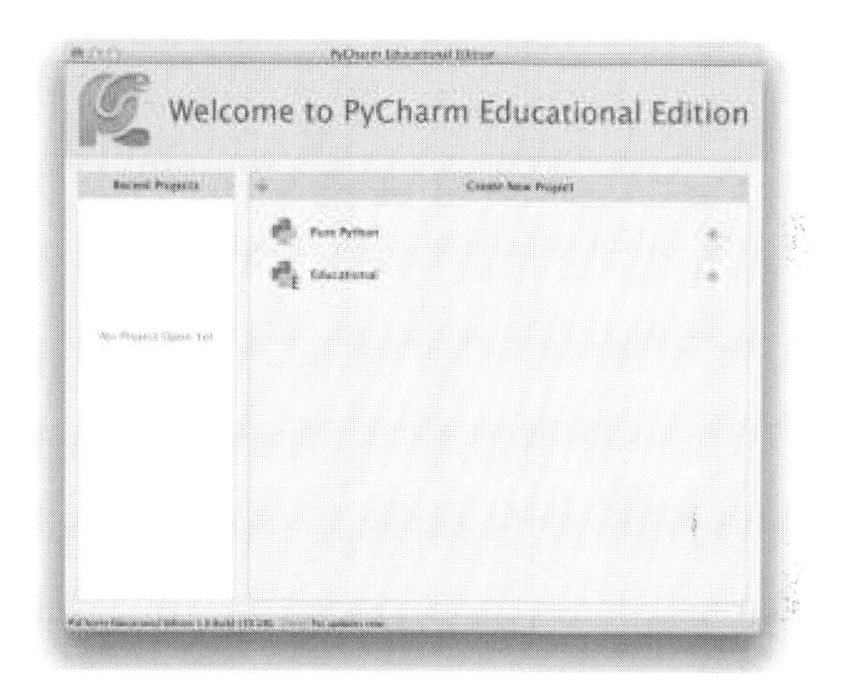

Select 'Pure Python'. A new screen will appear asking for the location of the file that you want to use. By default, it is named as 'Untitled' as shown in the image below. Change it to "Hello World" and click on the 'Create' button.

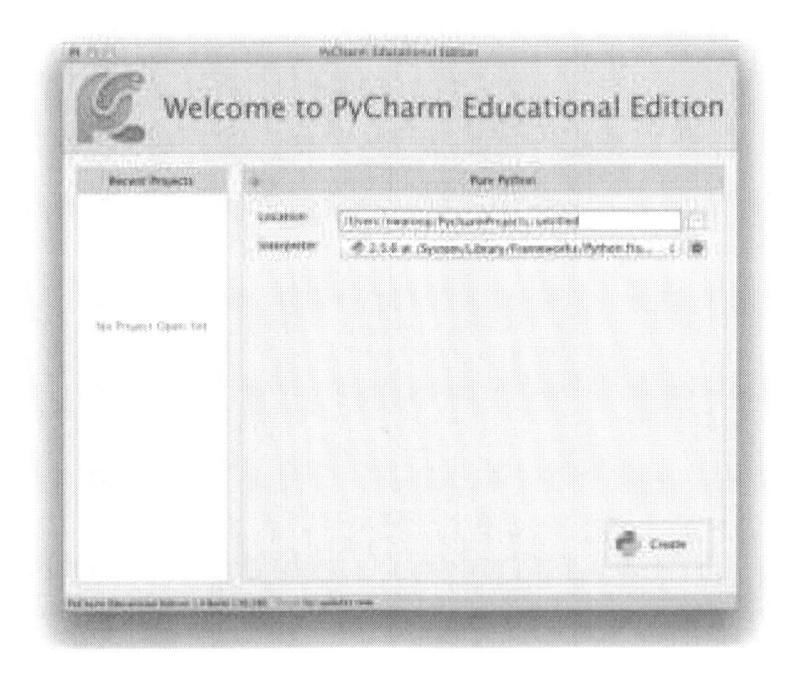

You will notice that a new project named Hello World will be displayed in the left sidebar. Right click on this project and go to 'New' and select 'Python File' as shown in the image below.

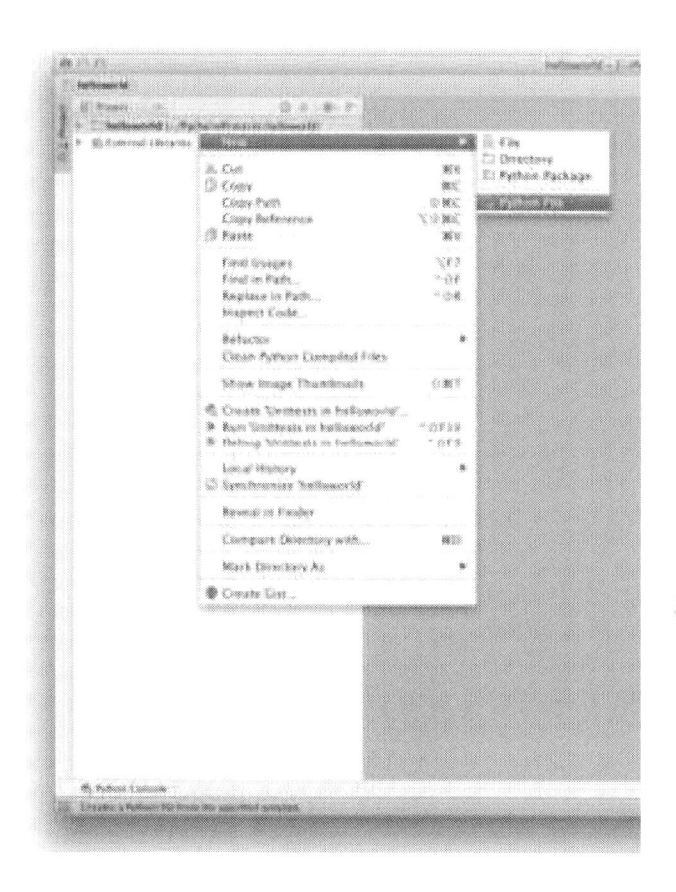

Now, a new dialogue box will appear on your screen asking for the filename. Simply type "hello" in it as shown in the image below.

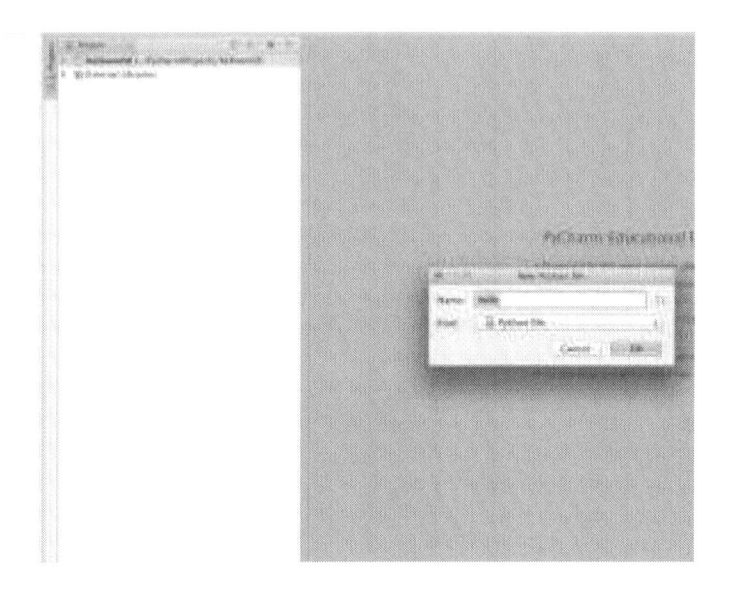

The file will now open and will be listed as hello.py in the left sidebar of the editor. Delete whatever is written by default in the program line 1, and then type –

Print("hello world")

Now, without selecting any text that you wrote, right click on the line and select Run 'hello' or press F10 from your keyboard.

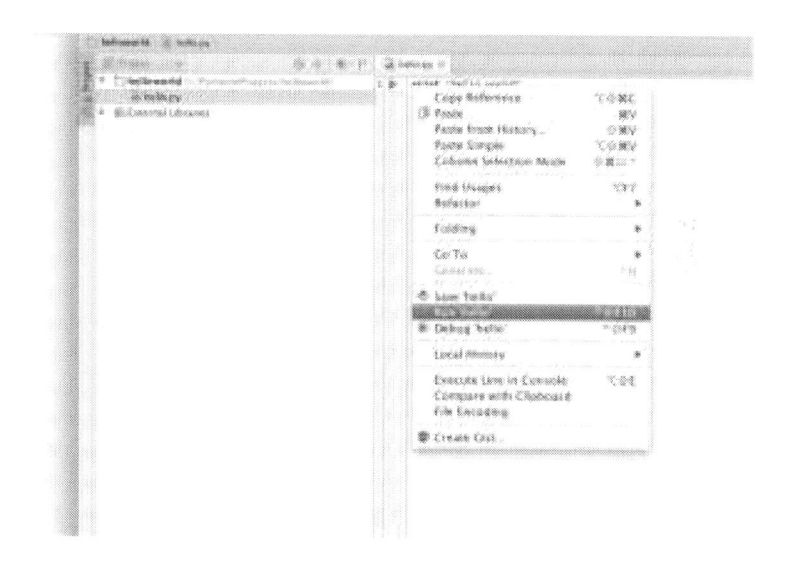

You will now see that PyCharm shows the result of your program just below it. Look at this picture to understand it better –

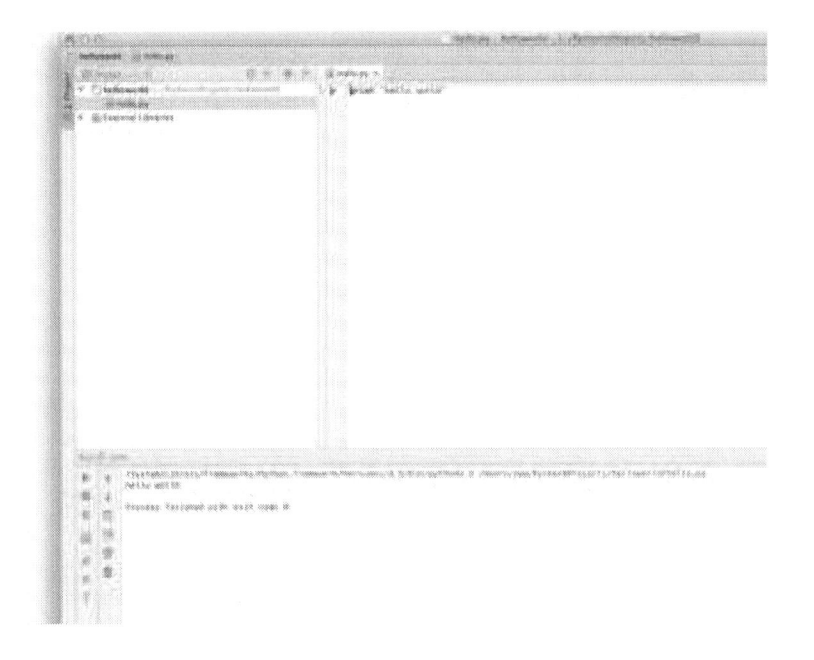

Saving your program is also easy. Simply right click and select Save 'hello'. To save all your programs, select a new location on your computer or create it. I am recommending the following locations for different operating systems –

/tmp/py on Mac OS X

/tmp/py on GNU/Linux

C:\\py on Windows

This was the standard method of using an editor to write and save your programs or projects in Python. This method is also known as using source files to write programs in Python. Now, let us try to run the program using the source file you just created and saved.

Using the Source File

In the previous step, you saved the file hello.py. To run this file, open the terminal and change the directory to that of your saved files on Python (that is: /tmp/py for

Linux and Mac OS X and py for Windows). Once you are in that directory, type the following command:

python hello.py

Here, hello.py is the file that you just saved. You will see that Python will instantly display the output of the program that is present in hello.py file.

If you could follow the process until now, then it is time to congratulate yourself, as you have written, run and saved your Python program, which is the biggest challenge for beginners learning the software.

Now, let's take it to the next step and learn a few more basics of Python before we start to write programs in it. See you in the next chapter.

Python Basics for Beginners

Of course, you just do not want to write print statement programs in Python. You would want to create more, I am sure. You would like to give Python some inputs that you can manipulate to get the desired output, don't you? But, you cannot do this as you don't know basic terms used in Python and their meanings. However, after reading this chapter, you will be able to quickly create a simple program as you wish. So, let's get started!

Comments

You will most often see a symbol # in python programs, this symbol is used for making comments. The text written next to this symbol is known as a comment. These are mainly used as a means of creating notes in the program for both developers and readers, so that they can know where they are, in the program. For example –

Print('I am Sushant' # This is a simple print statement of my name

In the above program, the comment tells the reader that what he is seeing is a simple print statement and nothing else. In

Python, the comments can be used for any of the following –

- Explaining the assumptions.
- Explaining the important decisions.
- Explaining the important details.
- Explaining the problems that you are trying to find answer for.
- Explaining the problems that you are trying to overcome, etc.

Literal Constants

Numbers such as 2,3,4,5, etc., and strings like 'This statement is a string' are examples of "literal constants". These are called literal constants because you can only use their value. The number 5 for example, will always represent nothing other than its value. This is because its value can never be changed, 5 will mean 5 no matter what the case is.

Quotes

You will see quotes being used in several programs that are written in Python, and you are advised to pay close attention to this section. There are three types of quotes used in Python; namely, single quote, double quote, and triple quote.

Quote

These can be used for specifying a string, example –

'This is a quote'

All white spaces i.e. tabs & spaces, within the single quotes, are saved as they are.

Double Quote

There are times when the single quote comes in between the sentence of the string, for example – *'What's that?'*

Python will now get confused as there are 3 single quotes. It will not understand which quote is the end, the second or the third. To tackle this problem, we use double quote. For example –

"What's that?"

Python will know that the string starts and ends with " and not ', and will work without getting confused.

Triple Quote

You can use the triple quote when you want to specify a multi string in your program, you can include both single quote and double quote freely inside triple quote. Below is an example for you to understand properly.

'''Here is a trip quite string and you are at first line.

You are now at the second line.

"What's that?," I asked.

He said "My car"

'''

The Format Method

When you are creating a program, sometimes you will come across a situation where you will need to build strings based on some other information. To do so, we use the "format" method. It is denoted by format().

Below is a program to help you understand this method clearly.

age = 23

name = 'Sushant'

print '{} was {} years old when he started sketching'.format(name, age)

print 'Why is {} still sketching?'.format(name)

The {} represents a value to be included, and which value to select is represented by the format. In the above case, there are 2 values in the first sentence and the format provided is - *format(name, age)*

This means that the first value is the name and the second value is the age. Python will now take these values from the information provided to create the output. The output of this program is –

$ python str_format.py

Sushant was 23 years old when he started sketching

Why is Sushant still sketching?

In the above output, str_format.py is the name that I used to create this file in the editor.

Escape Sequences

Let's go back to the single quote example again – 'What's that?'. We know that Python will get confused here, and to tackle this problem, we may use a double quote, or we may use an escape sequence. We need to tell python here that this single quote does not means the end of the string. We can do this by using "escape sequence" and it is denoted by \'. For using it in the above problem, all we have to do is write it in the following manner –

What\'s that?'

Let's move to another example where we will use an escape sequence. Suppose you want to write a string which contains two lines. One way to do so is by making the use of triple quote, and the other is by using an escape sequence - \n . This is used to indicate the start of a new line. An example is:

'You are at first line\nNow you are at the second'

Is same as -

'You are at the first line

Now you are at the second'

If in a string there is only a single backlash towards the end of line, then this is an indication for python that the same string continues in the next line without adding any new line. For example –

*"You are at the first statement. *

This is the same statement."

 Is same as -

"You are at the first statement. This is the same statement"

A Program to Practice

Below is a program for you to practice along with the explanation so that you can understand what's happening in the program. If you can understand this simple program, then you are ready to move to the next chapter of this book. If you are not able to, just read through this chapter again for a fuller understanding.

Type and run the following program on your machine –

Filename : sample.py

a = 6

print a

a = a + 1

print a

b = '''You are at the first line of multi string.

Now you are at the second line.'''

print b

The output of this program is –

6

7

You are at the first line of multi string.

Now you are at the second.

Explanation – Let's look at what we just did. First, by using the assignment operator (=), we assigned a value of 6 to the variable a (6 is literal constant here). After this, using the print statement which you are familiar with by now, we printed the value of a. Now, we add 1 to the initial value of a and stored it as a final value for our next step. We then printed the final value of a which is 7. In the same way, we created a string b and printed it on our screen.

Now, let's move to the next chapter of this book and learn some of the python operators which are often used.

Various Operators & Expressions Used in Python

In this chapter, we will learn various operators along with their precedence order in Python. Operators in Python are included in expressions that you write. Thus, it is important that you understand both the operators and expressions. For a start, you should understand that most statements written in Python contain expressions. One of the simplest examples of an expression is a mathematical operation. Let's take for example,

2+3

This is an example of an expression. The expression can be further broken down into operators and operands. Operands are some unique data on which an operator operates. In the above example, 2 and 3 are operands. Operators, on the other hand, are functionalities that do something and can be represented by symbols. In the above example, + is an operator, which adds the two operands. Let's now study the operators in detail.

Operators in Python

+ (plus) operator

This operator adds the two objects. For example –

3 + 10 = 13

In simple words, 'a' + 'b' gives 'ab'.

- (minus) operator

This operator subtracts the two objects. For example –

5 – 2 = 3

If the first operand is absent, then python takes it as 0 by default.

* (multiply) operator

This operator gives the result by multiplying the objects. For example –

2 * 3 = 6

When applied to strings, the operated repeats it for the given munber. For example –

'string' * 3 = 'string string string'

** (power) operator

This operator returns the a to the power of b. For example –

4 ** 4 = 4^4 = 256

/ (divide) operator

This operator divides the first value from the second. For example –

4 / 2 = 2

% (modulo) operator

This operator gives the remainder after dividing the first value from the second. For example –

13 / 3 = 1

<< (left shift) operator

This operator takes the number of bits specified in account and shifts the bits to left by that number. (Every decimal value is represented by binary digits i.e. 0 and 1 (or bits) in the memory)

4 << 2 gives 16

The decimal value 4 is represented by 100 in bits.

Left shifting the 4 by 2 bits gives 10000 which is equal to 16 in decimal.

>> (right shift) operator

This works the same way as the left shift, the only difference is that it makes a shift towards the right.

For example -

4 >> 1 gives the answer as 2.

The decimal value 4 is represented in bits by 100, when this value is right shifted by

1 bit, it gives us 10, which is represented by 2 in decimal value.

& (bit-wise AND) operator

This operator gives the Bit-wise AND result for the numbers. Example -

7 & 1 gives 1

| (bit-wise OR) operator

This gives the bitwise-OR result. Example -

7 | 1 gives 7

^ (bit-wise XOR) operator

It will give the Bitwise-XOR result. Example -

7 ^ 2 gives 5

~ (bit-wise invert) operator

This operator works on the bitwise inversion method. For a, the bit wise inversion is –(a+1) For example -

~6 gives -7

< (less than) operator

This is a comparison operator which tells if a is less than b or not. All comparison operators gives answer as *True* or *False*.

5 < 3 gives False and 3 < 5 gives True.

Comparisons done using this operator can also be done arbitrarily. For example –

1 < 3 < 9 gives True

> (greater than) operator

Another comparison operator which tells if a is greater than b or not.

Example: 9 > 1 gives the result as True

If both of the operands are numbers, then they are converted to a common type before applying the operation. Otherwise, this operator always gives the result as False.

(less than or equal to) operator

The operator looks like the symbol of comment, however, it works in a totally different manner. It tells us if the a is less than or equal to b.

a = 5; b = 9; a # b gives the result as True

>= (greater than or equal to) operator

This operator works in same manner as the # operator, the only difference is that it checks for greater than or equal to.

a = 6; b = 1; a >= 1 gives the result as True

== (equal to) operator

Another comparison operator in python which tells if the given numbers are equal or not. For example -

a = 3; b = 3; x == y is True

a = 'xyz'; b = 'xyZ'; x == y gives the result as False

a = 'xyz'; b = 'xyz'; x == y gives the result as True

!= (not equal to) operator

Another interesting comparison operator which checks if the two numbers are unequal or not. For example -

a = 5; b = 9; a != b gives the result True

We will now look at Boolean operations. These are kind of tricky, therefore pay close attention while reading them.

not (boolean NOT) operator

Per this operator, if a is True, then it returns the answer as False. If a is False, then it gives the answer as True. For example –

a = True; not a gives the answer as False

and (boolean AND) operator

In this operator, if a is False, then a and b are returned as False. Although if a is True, then it returns to evaluation of b. For example -

a = False; b = True; a and b gives the result as False, because a is False.

In the case, above, Python will not evaluate b. This is because it knows already that the value to the left of 'and' operator is False. Hence, it ignores the right-hand side value and simply displays the result irrespective of it. This is known as the short-circuit evaluation.

or (boolean OR) operator

For this operator, if a is True, then the answer is given as True, else if a is False, it will return for the evaluation of b. For example -

a = True; b = False; a or b gives the answer as True

Don't get confused in between and operator and or operator in python. Give yourself proper time to know how they work.

Operator Precedence Table for Python

Before moving on, let's first understand what precedence means. Consider an expression – 5 + 8 * 24.

Here, which operation will be executed first, the multiplication or the addition? As per the concepts that we studied in high school, the multiplication operator will be solved first. This means that the addition operator has lower precedence than the multiplication operator. Similarly, operators have different precedence level in Python. Below is a precedence table for python. In this table, the operator with lowest precedence is kept at the top while the operator with higher precedence is kept at the bottom. This means that the table is in increasing order of precedence for python operations.

Lambda	Lambda operator
if – else	Conditional expression
Or	Boolean-OR expression
And	Boolean-AND expression
not x	Boolean-NOT expression
in, not in, is, is not, <, #, >, >=, !=, ==	Comparisons, including membership tests and identity tests
|	Bit-wise-OR expression
^	Bit-wise-XOR expression
&	Bit-wise-AND expression
<<, >>	Left and right shift
+, -	Addition & subtraction expression
/, *, %, //	Division, Multiplication, Remainder and Floor Division expression
-x, +x, ~x	Negative, Positive, bit-wise-NOT expression
**	Exponentiation expression
x[index], x[index:index], x(arguments...), x.attribute	Subscription, slicing, call, attribute reference expression

Note - Objects with the same precedence are included in a single row separated by commas in the table above. You may see that some of these operators were not discussed before in this chapter. This is because they are a bit advanced and hence are not fit for beginners.

Expressions

We discussed previously what expressions are. To end this chapter, we have presented a program for you to do in Python on your machine. Save this program as expression.py. Then test yourself to see if you know the meaning of every line written in the program without looking at the explanations provided. However, I would recommend that you read this chapter again and check to ensure that you understand what variables, operators, and expressions are in the program even after reading the explanation.

Sample program –

Program for mathematical calculations

length = 10

breadth = 3

area = length * breadth

print 'The solution for area of rectangle is', area

print 'The perimeter for the same is', 2 * (length + breadth)

Output of this program is –

$ python sample.py

The solution for area of rectangle is 30

The perimeter for the same is 26

Explanation: We started by providing the values of breadth and length of the rectangle. These values are used in the program above to calculate the area and perimeter of the rectangle by using expressions in python. We created an expression for finding the area, which is later printed on the screen using along with the statement - ***The solution for area of rectangle is.*** After that, we directly found the value of perimeter by writing the expression inside the print command next to the line - ***The perimeter for the same is***. Python displays every result in a neat way, taking care of adding spaces by itself if we forget to add them ourselves. This makes the program easy to read and also makes the life of the programmer tension free as python itself takes care of small mistakes that a user might commit while writing the program.

Control Flow Statements in Python

So far, we have created programs in Python where a series of statements has been executed following from top to down order. However, let's say that you want to change the order that this execution follows. For example, you need to write a program which prints good evening or good afternoon or good morning based on the time. Here, you are asking your program to make some decisions. In Python, you can do so with the help of control flow statements. There are many control flow statements in Python, namely:

- If statement
- While statement
- For loop
- Continue statement

But, as this book is written as a beginner's approach to Python, we will only be discussing the first two statements followed by recursive functions in Python.

The If Statement

The "If" statement is used in python to check the satisfaction of a condition. If it is satisfied, then we execute a certain group of statements (this block is known as if-block). Otherwise, another group of statements is executed called else-block. This else-block is usually optional.

Some programmers use the if statement as an if-then statement, as the statement is executed when the condition is satisfied, and the else statement as an if-else statement as it is executed when the condition is not satisfied.

Let's learn it with the help of an example, write and save the following program in your python as if_statement.py

```
number = 25

input_guess = int(raw_input('Write one integer : '))

if input_guess == number:
 # Your new block is here
 print 'Congrats my friend, you made it to the end.'
 print '(but you did not bag any prize!)'
 # Your block ends
elif input_guess < number:
 # Another block
 print 'Nopes, try guessing higher than this'
else:
```

    print 'Nopes, try guessing lower than this'

print 'Done'

The DONE line is always executed once the if statement I executed.

The output that you will get when you will run this program is –

$ python if_statement.py

Write one integer: 43

'Nopes, try guessing lower than this

Done

$ python if_statement.py

Write one integer: 10

Nopes, try guessing higher than this

Done

$ python if_statement.py

Write one integer: 25

Congrats my friend, you made it to the end.

(but you did not bag any prize!)

Done

In the program created above, we first asked the user to enter his guess for the number and then we checked his input against the number we already have. Firstly, we set any integer for our variable number, like we set it to 25 in the above program. After that, we recorded the guesses made by the user using the raw input () function. For now, just remember that function is a reusable piece of a program. We then waited for the user's input after supplying a string to the raw_input function that prints it on the screen. Once the user has entered a value and hits the enter key on his keyboard, the raw_input () function takes the entered value as a string. Then with the help of int, this string is changed to an integer and after that it is stored as a value for variable guess.

After that, we compared the number with the value we already had. If the number is equal, a message is printed on the screen. You will notice that the "If" statement has a colon at its end, this way, we are telling Python about the group of statements that follows. Next, we checked if the input value is less than the value that we already have (that is 25). If it is so, a message is printed informing the user that they need to make a guess of higher value than that. You will notice that there is an elif clause being used in the program above. This clause combines the two if-else-if else statements and present them as one combined statement. This makes the program easier to understand and also reduces the indentation required. After that, we check if the input value is greater than the value that we already have. If so, then a message is displayed to the user asking him to guess a lower value than that. Always remember that the else and elif statements always have a colon towards the end, which is then followed by their respective group of statements.

Once Python has completed the execution of one if statement along with its elif and else clause, it then moves to the next if statement. However, we don't have another if statement in the program above. What we have in the next statement in the main block is a simple print 'Done' statement. After this, Python sees the end of the program and simply finishes up everything.

The While Statement

The control flow statement allows the user to repeatedly run a group of statements if the condition is satisfied. While statement sometimes is also known as looping statement and is used by many people to create recursive functions in Python. You can use this control flow statement with the if statement as shown in the example below.

Write and save this program in your python as while_statement.py.

number = 25

processing = True

while processing:

input_guess = int(raw_input('Write one integer : '))

if input_guess == number:

print 'Congrats my friend, you made it to the end.'

 # it will make the while loop stop

```
    processing = False

    elif input_guess < number:

    print 'Nopes, try guessing higher than this.'

     else:

    print 'Nopes, try guessing lower than this.'

else:

 print 'Congrats my friend, you made it to the end.'

 print 'Your while statement ends .'

 # You can write anything else from this point

 print 'Done'
```

Once you will run this program, your output will be like this –

$ python while_statement.py

Write one integer : 60

Nopes, try guessing lower than this.

Write one integer : 12

Nopes, try guessing higher than this.

Write one integer : 25

Congrats my friend, you made it to the end.

Your while statement ends.

Done

You must have noticed up to now that we are using the name guessing game as a way of explaining the while statement. The advantage here is that the program keeps running until the user has guessed the number correctly, hence, we don't have to run the program again and again for every guess like we did in the last section.

Before while loop in the program above, we set the processing variable to True. Then we first moved the if statements and the raw_input to inside of while loop. Later, we first ensured that the variable processing is True. Once it is verified, the while loop is executed. The condition is verified again after executing the while loop, which in the program above is the processing variable. If the condition is still satisfied, then the group of statements for while loop is again executed. If it is not, then we run the optional else group of statements and later we proceed to the next statement. When the while loop condition is not true, the else statement is executed by the program. The appropriate statement is displayed to the user depending on the guess he made. If the guess is correct, the while loop ends and the program terminates with a Done message printed on the screen.

The Recursive Function in Python

In Python, a function which repeats itself until its terminating condition has been satisfied is known as a recursive function. The while statement, and for loop statements are usually used to create such functions. The example we discussed above for the while statement is a recursive function, as it keeps on repeating itself until its terminating condition is satisfied. The same recursive function can also be created with the help of the if else statement, as shown in the example below –

Let's create a program to find the factorial of a number. We know that the n! in mathematics is n (n-1)! When n>1. Keeping this in mind, the program created to find the factorial in Python is as follows:

def factorial(n):

if n == 1:

return 1

else:

*return n * factorial(n-1)*

print factorial(3)

The output of this program will give the factorial value. If N = 1, then the result displayed will be 1, else the value N will be multiplied with the factorial N-1. The program will then continue N-1 till it has reached N=1.

However, if you try the program to find the factorial of 3000 instead of 3, you will get an error message. This is because when a recursive function is executed, Python stops the function automatically after 1000 repetitions to prevent system crash due to high memory usage by a recursive function.

Functions

In the example, above, you must have noticed the term def. This word is used to define a function in Python. Functions are nothing but a reusable piece of program. With functions, you can give a block of statements a name of your choice. This allows you to run that block of statements anywhere in the program using the specified word. This is the reason why the factorial word was mentioned again before (n-1) in the second to last line of the above program.

Files & Storing Persistently in Python

We have already learned how to write a program in Python, how to run it and how to save it. However, we can open and use the files for writing or reading by creating an object of the file and using its writing, reading or readline methods appropriately. The ability to write or read the file totally depends on the mode that you have specified for the file opening. Let's understand this with the help of a simple program.

```python
simple = '''\
Programming is not boring
When your are programming
if you want to make it fun:
 you should definitely use Python!
'''

# Opening the file for 'w'riting
f = open('simple.txt', 'w')
# Now you can write to this file easily
f.write(simple)
# We can close this file now
f.close()
```

```python
# If no mode is specified for the file,
# then 'r'ead mode is selected by default
f = open('simple.txt')
while True:
    line = f.readline()
    # EOF is indicated by the Zero length below
    if len(line) == 0:
        break
    # At the end of every line there is a newline
    # because it is reading from file.
    print line,
# Now lets close this file
```

f.close()

The output of this program is -

$ python io_using_file.py

Programming is not boring

When your are programming

if you want to make it fun:

you should definitely use Python!

Explanation: We start by simply opening a file and giving it a name io_using_file.py. The mode for opening the file can be the read mode ('r'), append mode ('a'), or write mode ('w'). We can also define if we are doing these operations (read, write, append) in binary mode ('b') or in text mode ('t'). If you open a file using open(), the file will be opened in text mode with read mode enabled. In the example, above, we first opened our file in write text mode and used the write method of the file object. After writing to the file, we closed it using the close function, close().

After that, we opened the file again for reading. As we didn't specify any mode, the text file mode is selected for reading by default. Then, we read each line of the program, until an empty line is returned by the program. An empty line means that we have reached the end of our file. At the end, we closed the file using the same close function again.

Storing Persistently in Python

Let us suppose that you have written an awesome program in Python and you spend your whole afternoon writing it. Now, you have to go for your dinner and you must close your Python as it is night already. You now will want to save the program so that you can open it again and use it to continue working. However, the save function only saves the file in text format (.txt) which is not optimal because once your program is written in a text file, it does not contain any of its original data structure. It is simply a normal text file. If you want to use this file to continue working, then you must process it and bring it back to its original data structure.

A solution for this problem is to save the Python data object of your work as itself. So, when you open your program later after dinner, you can just continue working on it. Python has a standard module for doing so, called Pickle with which you can store any plain Python object in a file and can continue working on it later. The method of storing the data is known as pickling and usually it has the file extension of .p or .pkl. Also, the method of loading the data is known as unpickling.

Pickling and unpickling involve the use of reading and writing routines that we discussed in the previous section of this chapter. Pickling is a module and it needs to be imported first before you can use it. Let's see how we can save the data for a simple program that contains some grades of a few students.

grades = {'Leonardo':75, 'Mona':98, 'Lisa':80, 'Nelson':65}

import pickle   # we need to import pickle first

```
f = open('grades.p', 'w')    # This creates a new pickle file

pickle.dump(grades, f)       # This will dump the whole data to f

f.close()
```

In the above program, we imported the pickle module first, then we created a pickle file with a .p extension where the data will be dumped. We created and opened this file with writing permission. After that, we used *pickle.dumb* to dump the data of a current program to f which is the pickle file.

Pickling was easy, wasn't it? In the same way, unpickling is also easy, just see the steps below to load the same program that we dumped before.

import pickle   #we need to import pickle again to use it

f = open('grades.p', 'r')   # 'r' here is for reading

mywork = pickle.load(f)   # this will load the data as mywork

f.close()

print mywork

 # prints {'Leonardo': 98, 'Mona': 75, 'Lisa': 80, 'Nelson': 65}

First, we imported the pickle module to use it. After that, we opened the pickle file with a ".p" extension that we created earlier with reading permission. Then we used the *pickle.load* module to load the data from that file as mywork. Then we used the print module to simply print all the data that we loaded.

cPickle

If you are trying to dump or store a large volume of data, then you will notice that the pickle module works at a very slow speed. An alternative available in Python to solve this problem is to use cPickle. This module works the same as Pickle, with the only difference being that it is almost 1000 times faster than the standard pickle module because it is written in C. To use this module, all you must do is write *import cPickle* instead of *import pickle*.

Sample Programs for Practice

Up to this point in the book, we have learned everything that a beginner needs to know about Python. It is now left for us to start writing some problem-solving programs. In this chapter, there is a total of 6 sample problems that have been provided for you to practice. These problems are basic yet effective to help you refine your knowledge of Python. Let's get started!

Sample Program #1

In this program, you will learn how to find the square root of the positive numbers.

Program Code –

```
# This is a program to find out the square root of the given number

# change this below value to any number you want

number = 25

# To take the input from the user, simply uncomment

#number = float(input('Enter the number to find square root: '))
```

*number_sqrt = number ** 0.5*

print('The solution of %0.3f is %0.3f'%(number ,number_sqrt))

The program above will give you the square root of any positive number that you want. Change the value of 8 in the program above to a number of your choice and see if you are able to run the program successfully.

Sample Program #2

In this program, you will learn how to find the area of a triangle with the help of Python.

Program Code –

This is a program to find the area of a triangle easily

a = 7

b = 5

c = 6

To take inputs from the use, simply uncomment

#a = float(input('Enter the measurement for side1: '))

#b = float(input('Enter the measurement of side2: '))

#c = float(input('Enter the measurement of side3: '))

```
# now find the semi-perimeter
p = (a + b + c) / 2
# now find the area of triangle
area = (p*(p-a)*(p-b)*(p-c)) ** 0.5
print('Area of the given triangle is %0.2f' %area)
```

In this program, a, b, and c represent the 3 sides of the triangle. We first calculated the semi perimeter of the triangle and later used that value to calculate the area of the triangle. A little bit of mathematics knowledge is a great help in creating such programs.

Sample Program #3

In this program, you will learn how to use your Python to convert kilometers to miles.

Program Code –

```
# The kilometers to miles converter
# change the value below to test
kilometers = 6.5
# below is the conversion factor
conversion_factor = 0.621371
# now calculate the miles
miles = kilometers * conversion_factor
```

print('%0.3f kilometers is equal to %0.3f miles' %(kilometers,miles))

You can change the value of kilometers to your desired value to test this program.

Sample Program #4

In this program, you will learn how to easily differentiate between odd and even numbers using your Python.

Program Code –

 # Program to find if the number is even or odd

if division by 2 give a remainder of 0 the number is even.

The number is odd if remainder is odd.

number = int(input("Enter a number to test: "))

if (number % 2) == 0:

   print("{0} is an even number".format(number))

else:

   print("{0} is the odd number".format(number))

In this question, instead of using the values by default, we asked the user to add the values that he wants to check for even and odd numbers.

Sample Program #5

In this program, you will learn how to find the LCM of given numbers using your Python.

Program Code –

Python Program to find the L.C.M. of two input number

start by defining a function for this program

def L_C_M(m, n):

  """This defined function gives L.C.M

for the two numbers"""

select the number which is greater

if m > n:

    greater = m

else:

    greater = n

while(True):

    if((greater % m == 0) and (greater % n == 0)):

        L_C_M = greater

        break

    greater += 1

```python
    return L_C_M

# change the values of numbers below for testing

number1 = 24

number2 = 54

# uncomment the following lines to take input from the user

#number1 = int(input("Enter the first value for testing: "))

#number2 = int(input("Enter the second value for testing: "))

print("The L.C.M. of", number1,"and", number2,"is", L_C_M(number1, number2))
```

You can simply change the values of two numbers in order to test the program for different numbers.

Sample Problem #6

In this program, you will learn how to solve a quadratic equation using your Python.

Program Code –

*# quadratic equation is of the form px**2 + qx + r = 0*

 # start by importing cmath module as shown below

import cmath

p = 1

```python
q = -2

r = -3

# To take coefficient input from the users

# p = float(input('Enter p: '))

# q = float(input('Enter q: '))

# r = float(input('Enter r: '))

# now find the discriminant

a = (q**2) - (4*p*r)

# calculate two solutions

solution1 = (-q-cmath.sqrt(a))/(2*p)

solution2 = (-q+cmath.sqrt(a))/(2*p)
```

print('The solutionmfor the given equation are {0} and {1}'.format(solution1,solution2))

In this problem, we used the cmath module to create this program. First, we calculated the discriminant and then we found the roots of the given problem.

Exercise for You

The above programs were simple but the last two had few terms that we did not discuss in this book. Find out what those terms mean and try to make a new program using those terms just like in the examples provided for you above.

Conclusion

This book was written as a beginner's approach to Python programming. It discussed everything that a beginner needs to understand to create his first simple program using the software. I hope that this book was valuable to you. Always remember that you will learn Python more easily if you use it simultaneously while reading this book. I hope that you have fun in learning your first programming language. Lastly, I want to say thank you for downloading this book.

Happy learning!

Printed in Great Britain
by Amazon